ENVIRONMENTAL RISKS AND AGRICULTURAL BENEFITS OF SOIL AMENDED WITH ANAEROBIC DIGESTATE.

BY

AGBOR KINGSLEY CHUKWUEBUKA

DEDICATION

This seminar paper is dedicated to the Almighty God for his Compassion and Grace in my life.

ACKNOWLEDGEMENT

I am grateful to God for His Grace to make this work a reality. I also want to thank my supervisor Dr. Ihuoma N. Anyanwu, co-supervisor Mr. S.T. Nwoba, HOD Dr. Okoro Nworie, seminar coordinator Dr. Mrs. Ogochukwu C. and all the lecturers of BIO/MCB/BTG Federal University Ndufu-Alike Ikwo (FUNAI) for equipping me academically, and making me competent in my field of study.

TABLE OF CONTENT

Title page………………………………………………………………………………i

Dedication……………………………………………………………………………..ii

Acknowledgments……………………………………………………………………iii

Preface………………………………………………………………………………...iv

Table of contents………………………………………………………………………v

List of tables…………………………………………………………………………..vi

List of figures………………………………………………………………………...vii

List of abbreviation…………………………………………………………………viii

Abstract……………………………………………………………………………….ix

CHAPTER 1

1.0 Introduction………………………………………………………………………1

1.1 Origin of Digestate………………………………………………………………3

CHAPTER 2

2.0 Literature Review………………………………………………………………...5

2.1 Anaerobic Digestate……………………………………………………………...5

2.2 Different Types of Waste and the Digestate produced…………………………...13

2.3 Works Done on Digestate………………………………………………………………....15

2.4 Effects of Digestate……………………………………………………………………….16

2.4.1 Agricultural Effects of Digestate……………………………………………………….16

2.4.2 Environmental Effects of Digestate ……………………………………………………..17

2.5 Environmental Risks Associated with the Application of Anaerobic Digestate………...21

2.5.1 Risks of Nutrient Pollution……………………………………………………………….21

2.5.2 Risks of Soil Contamination……………………………………………………………...22

2.5.3.1 Physical Contamination………………………………………………………………...22

2.5.3.2 Chemical Contamination……………………………………………………………….22

2.5.3.3 Biological Contamination……………………………………………………………….23

2.6 Other Risks of Digestate to the Environment……………………………………………..23

CHAPTER 3

3.1 Conclusion ……………………..…………………………………………………..24

3.2 Recommendation ………………………………………………………………………24

3.3 Future Research …………………………………………………………...…………25

3.4 References………………………………………………………………………..26-31

LIST OF TABLES

Table 1. Characteristics of digestate.

Table 2. Example analyses for two types of the whole digestate.

Table 3. Chemical and agrochemical characteristics of digestate from vegetative experiments.

Table 4. Nutrient composition of Digestate.

Table 5. Example from Northern Ireland of the average nutrient composition over 52 weeks of feedstock (dairy cow slurry) and digestate in a mesophilic digester at the Agri-Food and Biosciences Institute

LIST OF FIGURE(S)

Figure 1. Flow Chart of Anaerobic Digestate.

LIST OF ABBREVIATION

AD - Anaerobic Digestion

DM – Dry Matter

FM – Fresh Matter

FW – Food Waste

GHG – Green House Gases

OM – Organic Matter

SOM – Soil Organic Matter

TS – Total Solids

ABSTRACT

Anaerobic Digestate is a by-product of anaerobic digestion and can be used as a fertilizer. Few studies on digestate and its physicochemical properties have shown that the amendment of soil with anaerobic digestate represents an alternative for sustainable agricultural production and/or influences soil fertility. However, these focused on soil organic matter content, soil pH, soil carbon, and nitrogen. Apart from the benefits, there may be environmental risks/impacts on soil biota such as plants, earthworms, and microbes. Thus, studies should explore the impact of anaerobic digestate on plants and ecosystem engineers (microbes, earthworms, enchytraeids, and collembolans) to ascertain its agricultural benefits and/or environmental risks.

CHAPTER 1

1.0 Introduction

Digestate is a nutrient-rich substance produced by anaerobic digestion that can be used as a fertilizer (NNFCC, 2017). It consists of leftover indigestible material and dead Microorganisms, the volume of digestate will be around 90-95% of what was fed into the digester (NNFCC, 2017). Digestate is not compost, although it has some similar characteristics. Compost is produced by aerobic micro-organisms, meaning they require oxygen from the air (NNFCC, 2017).

Digestate is the by-product of methane and heat production in a biogas plant, coming from organic wastes. Depending on the biogas technology, the digestate could be a solid or liquid material (Makádi *et al.*, 2012).

It contains a high proportion of mineral nitrogen (N) especially in the form of ammonium which is available for plants (Makádi *et al.*, 2012). Moreover, it contains other macros- and microelements necessary for plant growth (Makádi *et al.*, 2012). Thus the digestate can be a useful source of plant nutrients or an effective fertilizer for agricultural plants (Makádi *et al.*, 2012). On the other hand, the organic fractions of digestate can contribute to soil organic matter (SOM) turnover, influencing the biological, chemical, and physical soil characteristics as a soil amendment (Makádi *et al.*, 2012).

The digestate may be a very useful organic fertilizer that can be used to balance the financial as well as the environmental costs associated with the use of mineral fertilizer (Lukehurst *et al.*, 2010).

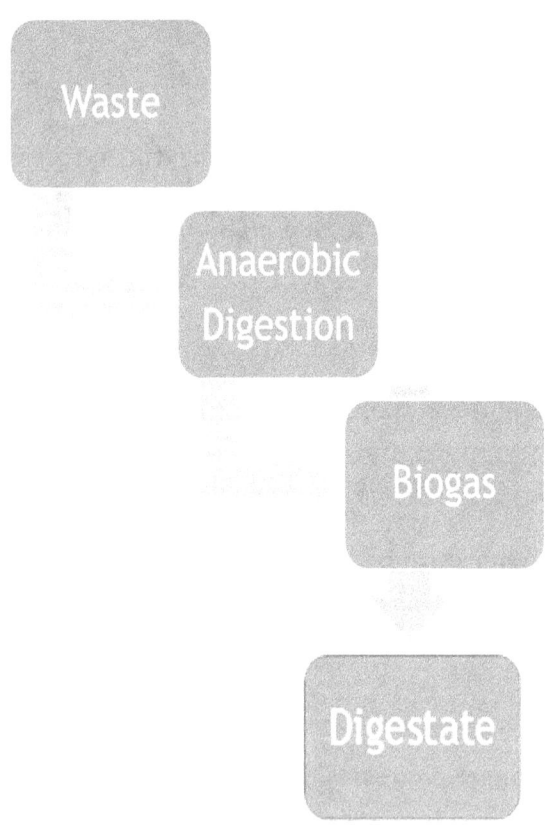

Fig. 1. Flow Chart of Anaerobic Digestate

The fertilizer value of a digestate depends on the nutrients present in the feedstock. However, digestate is the result of a living process and therefore has characteristics that are specific to each digester tank. Wastes from stables, agricultural residues, and food industry, municipal wastes, and dedicated energy crops are the main feedstock for anaerobic digestate in biogas plants (Möller *et al.*, 2012). If well handled, anaerobic digestate can be a valuable resource for agricultural production and a source of nutrients for soil fertility (Holm-Nielsen *et al.*, 2009).

Digestate can be obtained from vegetable waste, pawpaw, watermelon, pineapples, oranges, plantain, cassava, and banana peels, chicken and cow dung are the main feedstock for anaerobic digestate in biogas plants (Möller *et al.*, 2012).

There is a wide range of anaerobic digestates whose composition and aspect depend upon the type of biomass (feedstock) used and the configuration of the digester. Thus, spectroscopic techniques have recently demonstrated that anaerobic digestates inherit the chemical attributes of the feedstock from which they are produced (Provenzano *et al.*, 2011). Various types of feedstock and combinations of feedstocks such as, cattle manure (Gomez *et al.*, 2007), livestock manure and agricultural residues (Amon *et al.*, 2007; Tambone *et al.*, 2010), organic solid wastes, and sewage sludge (Gomez et al., 2007; Murto *et al.*, 2004), dairy manure and biowastes, food wastes and landscape wastes (Drennan and Distefano, 2010) and potato and sisal pulp wastes (Parawira *et al.*, 2004; Mshandete *et al.*, 2005) have been reported.

1.1 Origin of Digestate

Globally, historical evidence indicates that AD is one of the oldest technologies. Even around 3000 BC, the Sumerians practiced anaerobic cleansing of waste (Deublein and Steinhauser, 2008). However, the industrialization of anaerobic digestion began in 1859 with the first AD plant sited in Bombay (India). In 1897, an anaerobic digester at Matunga Leper Asylum in Bombay used human waste to generate biogas (Khanal, 2008). According to Deublein and Steinhauser (2008), other countries that pioneered the evolution of biogas technology were: France, in 1987 the streets lamps of Exeter amounts ($\leq 3\%$) of impurities such as hydrogen sulfide, ammonia, carbon monoxide, and other gases started running on biogas produced from wastewater; (Monnet, 2003). China, rural biogas system developed in 1920, while the national program started in 1958; in Germany, agricultural products were used to produce biogas in 1945. Today, China is credited as having the largest biogas program in the world with over 20 million biogas plants installed (Tatlidil *et al.*, 2009).

In Africa, anaerobic digestion technology was introduced in Africa between 1930 and 1940, when Ducellier and Isman started building simple biogas machines in Algeria to supply farmhouses with energy. Despite this early start in Africa, the development of large-scale biogas technology is still in its embryonic stage in this region, though with a lot of potentials (Ngumah *et al.*, 2013).

In Nigeria, the status of anaerobic digestion technology remains abysmal. The earliest record of anaerobic digestion technology in Nigeria was in the 1980s, when a simple anaerobic plant that could produce 425 liters of biogas per day was built at Usman Danfodiyo University, Sokoto (Dangogo and Fernado, 1986). About 21 pilot demonstration plants with a capacity range of between 10m3-20m3 have been cited in different parts of the country (Ngumah *et al.*, 2013).

Despite the many studies on anaerobic digestion, the benefits and/or impact of anaerobic digestate on the environment/agriculture is still fragmented in literature. This review explores the environmental risks and agricultural benefits of soil amended with anaerobic digestates.

CHAPTER 2

2.0 LITERATURE REVIEW

2.1 Anaerobic Digestate

Digestate is an excellent bio fertilizer that can help farmers to reduce their costs. (Nicola, 2012). It is an excellent alternative to chemically produced inorganic fertilizers. Using it improves the sustainability of farming by reducing emissions of greenhouse gases associated with fertilizer manufacture, and by reconnecting nutrient cycles (WRAP, 2017). Some digestates are made from livestock slurries, plant wastes, and food wastes such as dairy produce, meat, and fish. Digestate typically comes in three forms: whole (similar in its appearance to a livestock slurry), liquor, and fiber (WRAP, 2017).

2.1.1 Types of Digestate

Digestate, based on its physical properties, can be classified as whole digestate, liquid fraction or liquor, and solid fraction or cake, and their characteristics rely basically on the solid-liquid separation technique applied (Vilanova and Noche, 2016).

Also, digestate can also be classified, depending on the source of feedstock, such as agriculture-based digestate (manure and crops), digestate from food and municipal waste, and digestate from wastewater treatment plant (Vilanova and Noche, 2016).

Table 1: Digestate Characteristics (Al Seadi, 2004).

PARAMETERS	ABSOLUTE VALUES	CHANGE [a]
DM (%)	1.5–13.2	-1.5 to -5.5
Organic DM (%DM)	63.8–75.0	-5 to -15
Total N (%DM)	3.1–14.0%	[b]
Total N (kg mg -1 FM)	1.2–9.10	≈0
Total NH_4^+ (kg mg -1 FM)	1.5–6.8	?
NH_4^+ share on total N (%)	44–81%	+10 to +33
Total C content (%DM)	36.0–45.0	-2 to -3
C:N ratio	3.0–8.5	-3 to -5
Total P content (%DM)	0.6–1.7	[b]
Total P (kg mg -1 FM)	0.4–2.6	≈0
Water soluble P (% of total P)	25–45	-20 to -47
Total K (%DM)	1.9–4.3	[b]
Total K (kg mg-1 FM)	1.2–11.5	≈0
Total mg (kg mg-1 FM)	0.3–0.7	≈0
Total Ca (kg mg-1 FM)	1.0–2.3	≈0
Total S (kg mg-1 FM)	0.2–0.4	?
pH	7.3–9.0	+0.5 to + 2 units

In the table above, DM = Dry matter; FM = Fresh matter; mg=Milligram, kg = Kilogram

2.2 Wastes, Digestate Produced, and their physicochemical properties

Table 2: Example analyses for two types of the whole digestate (WRAP, 2017).

Parameters	Units	Food-based digestate	Manure-based digestate
Dry Matter (solids)	%	4.33	8.22
pH		8.41	8.22
Specific Gravity (density)	kg/m^3	0.99	0.97
Total N	kg/m^3	7.35	4.40
Readily Available N	kg/m^3	5.94	2.55
Total Phosphate (P$_2$O$_5$)	kg/m^3	0.48	1.35
Total Potash (K$_2$O)	kg/m^3	1.81	3.49
Total Magnesium (MgO)	kg/m^3	0.06	0.74
Total Sulphur (SO$_3$)	kg/m^3	0.44	1.28

Table 3: Chemical and agrochemical characteristics of digestate from vegetative experiments (Nicholas et al., 2015)

Elements	70:30 pig manure: Markets waste	Element	70:30 pig manure: Markets waste
pH-H$_2$O	7,62	Mobile P %	0,54

Moisture %	98,9	Mobile K %	1,25
Dry residue %	1,1	S (as SO$_4$) %	0,1
Organic C %	24,88	As mg/kg	<5,0
Total P$_2$O$_5$ %	7,67	Cd mg/kg	<1,0
Total N %	10,8	Cr mg/kg	13
Total K$_2$O %	9,02	Ni mg/kg	26
Total CaO %	7,6	Cu mg/kg	411
Total MgO %	2,89	Zn mg/kg	1409
Mobile N-NH$_4$ %	5,48	Pb mg/kg	8
Mobile N-NO$_3$ %	0,53	Hg mg/kg	<1

2.3 Some studies on anaerobic Digestate:

In the world, anaerobic digestion was used but primarily as a process for treating high-COD waste rather than as a means of generating energy (biogas). By the mid-1950s, France had over 1,000 anaerobic installations in various farm operations, which varied from simple covered tanks to complex digestion systems (Lesage and Abiet, 1952). In West Germany, this technology reached its peak in 1944–1945; the press gave wide coverage to the idea of using agricultural wastes in this process as feed and also about the development of different types of anaerobic plants. Since 1975, a number of these countries, in particular, Southeast Asian countries, have begun to give a thrust at the government level to exploit the potential of anaerobic digestion (Abbasi, 2012).

In Japan, anaerobic digestion has received considerable attention during the last few years from the point of view of pollution control, and for the treatment of livestock, industrial, and urban waste. Japan is the only country in the region that has adopted thermophilic digestion of some wastes (Abbasi, 2012).

In the USA, Canada, and Western Europe anaerobic digestion has been used mainly for processing animal manure till the mid-1970s. The advancements in high rate anaerobic digesters began with the introduction of the anaerobic filter in 1967 (Abbasi, 2012). It was followed by the introduction, one after another, of several other forms of anaerobic digesters capable of treating a wide variety of biodegradable wastewaters. Developed countries have given the initial thrust towards wastewater treatment using anaerobic digesters and it is being increasingly followed all over the world (Abbasi, 2012).

In Africa, a large number of digesters began to be installed in countries such as Indonesia, Papua New Guinea, Egypt, Uganda, Tanzania, Ethiopia, Zambia, Nigeria, and many other African countries. (Van-Brakel, 1980; Abbasi, 2012)

In Nigeria however, anaerobic digestate has remained at the level of institutional research work and pilot schemes. It's progress being stunted by ignorance, research at universities frequently considered as being too academic, lack of political will, and lack of an adequate coordinating framework (Ngumah *et al.*, 2013)

a) Nutrient Content of Digestate.

The anaerobic effluent of digestate contains macronutrients (N, P, K, Ca, S and Mg) and micronutrients (B, Cl, Mn, Fe, Zn, Cu, Mo, and Ni). Usually, its characteristics depend on the input material, operating conditions of the AD process, and digestate processing techniques (Lukehurst *et al.*, 2010).

The anaerobic digestate is rich in nitrogen (N), phosphorous (P), and potassium (K). After solid-liquid separation, the liquid part contains a high N percentage and the solid part contains high P content. In addition, the presence of heavy metals (Cd, Cr, Pb, Ni, Hg, Cu, and Zn) and organic pollutants can be found (Bustamante *et al.*, 2013).

All the nitrogen, phosphorous and potassium present in the feedstock will remain in the digestate as none is present in the biogas. However, the nutrients are considerably more available than in raw slurry, meaning it is easier for plants to make use of the nutrients (NNFCC, 2017).

The exact composition of digestate is determined by the plant's diet. However, typical values for nutrients are:

- Nitrogen: 2.3 - 4.2 kg/ton

- Phosphorous: 0.2 - 1.5 kg/ton

- Potassium: 1.3 - 5.2 kg/ton (http://www.biogas-info.co.uk/about/digestate/).

High proportions of P (phosphorus) and K (potassium) in animal diets are also excreted. Animal manures and slurries are therefore rich in plant nutrients. This is also the case for many other types of AD feedstock, making digestate a valuable biofertilizer. By making the best possible use of digestate as a biofertilizer, nutrients are returned to the land through natural cycles to replace the input of inorganic fertilizer. Recycling in this way closes a loop to create more sustainable agricultural production systems (NNFCC, 2017).

Other factors that affect the composition of digestate include the waste, as well as geographical and climatic conditions.

b) Fertilizer Value of Digestate

Digestate is an easy product to handle and apply and can be used successfully as a substitute for mineral fertilizers. The fertilizer value of digestate depends on the nutrients present in the feedstock. Anaerobic digestate draws carbon, hydrogen, and oxygen from the feedstock (http://www.res2.agr.ca/initiatiatives/manurenet/download/biogas). The most important advantages of organic fertilizer are their participation in the natural nutrient cycle, while inorganic fertilizer is added to it (http://www.websrv5.sdu.dk/bio/pdf/rap2.pdf). The utilization of digestate as a bio-fertilizer benefit both farmers and the environment. In the first place, it helps restore the natural recycling process of valuable and scarce nutrients, such as phosphorus. By spreading digestate on soil, the phosphorus contained in the biodegradable AD feedstock is brought back into nature to be incorporated into new organisms and to continue its cycle (Envirotech, 2017).

According to Aniche (2013), digestate has improved fertilizer efficiency due to its:

- homogeneity
- higher nutrient availability
- nitrogen content
- phosphorus content
- Potassium content
- micronutrients.

b) Physical, Chemical, and Biological Content of Digestate

A broad range of nutrients is contained in digestate including, among others, nitrogen (N), phosphorus (P), potassium (K), magnesium (Mg), and sulfur (S) (Nicholas et al., 2015). The fertilizer value of digestate depends on the nutrient value of the used feedstock. Since almost all macro-and micro-nutrients are conserved during anaerobic digestion, high-quality digestate can be ensured by feeding AD plants with high-quality substrates, such as source-separated organic waste. Irrespective of this, the content of the digestate strongly varies between AD plants and batches from the same digester. Macro-nutrients are vital for both plants and animals, however, animals do not utilize all nutrients, high proportions of nitrogen, potassium, and phosphorus are excreted. The resultant manure and animal slurry thus constitute a highly valuable substrate (http://www.envirotech-online.com).

- **Chemical content**

When organic material is anaerobically digested, most of the nutrients in the feed material are retained in the process and hence end up in the digestate (Debosz *et al.*, 2002). The digestate contains most of the macronutrients (N, P, K, Ca, S and Mg) and micronutrients (B, Cl, Mn, Fe, Zn, Cu, Mo, and Ni) needed by plants.

- **Microbial content**

Digestate is a living material and contains a wide variety of different microorganisms. For example, microorganisms from the preceding AD can be found in the digestate. These microorganisms are still active during the storage of the digestate, as shown by the post-production of biogas. Moreover, there is a risk of organisms not needed for the biogas process *per se* being present, among which are potentially pathogenic bacteria and fungi (Sahlstrom *et al.*, 2008; Schnurer and Schnurer, 2006; Bagge *et al.*, 2005; Sahlstrom, 2003). Different types of these organisms can pose a risk not only to the humans handling the digestate but also to the soil and plant system and to animals grazing on digestate-fertilized fields.

Pathogenic bacteria such as *Listeria, Salmonella, and Escherichia coli, Mycobacterium, Clostridium, Campylobacter* and *Yersinia* have been found in different substrates such as farm and slaughterhouse wastes and wastes from food processing industries (Sahlstrom, 2003). Spore-forming *Clostridia* and fungal spores have also been detected in digestate (Schnurer and Schnurer, 2006; Bagge *et al.*, 2005).

Table 4: Nutrient composition of Digestate (Anna, 2015)

Nutrients	Liquid 1	Available Kgs/m3	Liquid 2	Available Kgs/m3	Solid	Available Kgs/ton
Dry Matter (%)	2.8		4.0	40	26.3	263
Total Nitrogen (%)	0.44	2.2	0.36	1.8	0.75	2.6
Ammonium-N (%)	3085		2062		2435	
Total Phosphorus (%)	0.07	1.2	0.04	0.74	0.36	6.6
Total Potassium (%)	0.07	0.72	0.27	2.9	0.45	4.9
Magnesium (ppm)	100	0.1	500	0.5	2700	2.7
Sulphur (%)	335	0.34	284	0.3	1185	1.2
Calcium (ppm)	1250	1.25	1,400	1.4	8100	8.1

Zinc (ppm)	10	0.01	15	0.02	51.3	0.05
pH	8.0		7.8		8.2	
C:N ratio	2:1		4:1		15:1	
Organic matter (%)	2	20	2.7	27	20.8	208
Conductivity (EC) mg/cm	17.7		16.4		4.72	
Sodium (ppm)	700	0.7	900	0.9	1400	1.4

Table 5: An example from Northern Ireland of the average nutrient composition over 52 weeks of feedstock (dairy cow slurry) and digestate in a mesophilic digester at the Agri-Food and Biosciences Institute (Frost and Gilkison, 2010).

	Dry matter (g/kg)	Total N (g/kg fresh)	NH_4-N (g/kg fresh)	NH_4-N (% Total N)	pH
Feedstock	72.2	3.5	2.0	67.0	7.4
Digestate	59.3	3.6	2.4	80.5	7.9
Change	-17.9%	+2.8%	+20		
Standard deviation feedstock	8.50	0.52	0.36		0.34

Standard deviation digestate	5.22	0.48	0.43		0.23

2.1.2 Uses of anaerobic digestate

There is a wide range of utilization of digestate, which depends on the quality and the origin of the input substrate, as well as the type and characteristics of digestate. The most common use is, land application such as fertilizer and soil conditioner (Kratzeisen *et al.*, 2010) only if the heavy metal content (Cd, Cr, Pb, Ni, Hg, Cu, Zn) and organic pollutants make it suitable for agriculture use (Vilanova and Noche, 2016). Moreover, digestate can also be converted into compost (Fuchs et al., 2013), used for growing medium for plants and for land regeneration. Other studies proved uses such as solid fuel as a promising alternative after its drying and pelletizing (Wager-Baumann, 2011). Digestate can also be used as a building material. In addition, after the separation of the digestate, the liquid phase may have different end-uses. It can be spread directly to the land as N-rich fertilizer, or recirculated to the AD process as process water, or further treated to obtain concentrates or pure water (Bioenergy, 2017).

2.4 Effects of anaerobic Digestates in agriculture:

Anaerobic digestate can have different environmental and agricultural impacts.

a) To Ecosystem Engineers (earthworms, microbes)

In agricultural ecosystems, earthworms are often the major component of soil macrofauna. In these systems, earthworms and physical and chemical soil properties are important. They affect soil porosity and structural stability as well as soil organic matter dynamics and nutrient release. (Rossi *et al,*). According to Rachel *et al.* (2017), the ash-digestate blends caused earthworms mortality when applied at 340 kgN.ha^{-1}. This was potentially linked to changes in soil pH and elements speciation. But there has not been any study yet, on the effects of pure digestate on earthworms.

c) Plants

Digestate can enhance soil fertility and productivity, improving the plant nutrient status for potentially limiting nutrients such as N, P, K as well as for several micronutrients (Liu et al., 2009) which can be easily absorbed by plants. The absorbed nutrients from the digestate can improve the yield of the plant. However, there are relatively few studies that have tested the effect of AD on the survival rate of pathogens that affect plants. While plant pathogens can be treated with fungicides, many farmers try to avoid their use due to the expense and environmental concerns (Kajsa, 2015).

c) Soil Organic Matter Content

Digestate is a very complex material, therefore, its use has an effect on the wide range of physical, chemical, and biological properties of the soil, depending on the soil types (Makádi et al., 2008). The recycled organic wastes are suitable for contribution to maintain the soil nutrient levels and soil fertility (Makádi et al., 2008). Additionally, digestate is rich in organic matter and therefore it is a valuable humus producer. Usually, more than 50% of dry matter content is organic matter, which represents the basis for humus production (Tambone et al., 2007).

d) Soil pH

Digestate might contain various acidic compounds (e.g. gallic acid). The polycondensation, connection to organic and inorganic colloids, and transformation of these acids can have an effect also on the soil chemical properties and decrease soil pH (Tombácz et al., 1998), more particularly at the soils with high organic and inorganic colloid contents. Thus, regular monitoring of soil pH is necessary in the case of long-term digestate application. (Mariana *et al.*, 2012).

e) To Soil Carbon and Nitrogen

The influence of digestate on the mineralized nitrogen and carbon content in soil may depend, beyond the quantity of available nitrogen, and also on the microbiological activity of the compost (Jacques *et al.*, 2008). Normally, digestates contain a high amount of mineralized nitrogen, (as ammonia), which contains relatively low quantities in the form of lignin-rich materials. Therefore, nitrogen immobilization is not expected after the utilization of such products. The reason for the immobilization of nitrogen in soil by some digestates is that these products are not used fresh, but after an inadequate subsequent treatment, during which the digestate has been dry and has lost all the ammonia (Jacques *et al.*, 2008).

f) Other Soil Nutrients/Constituents

One of the main problems of digestate (and other N fertilizer) application is N leaching. However, Renger and Wessolek (1992) and Knudsen *et al.* (2006) found that the N leaching was dependent on the use of cover crops. Similar results were reported by Möller and Stinner (2009) who did not find differences in the soil mineral N content among different manuring systems in the case of winter wheat, rye, and spelled in autumn, before use of cover crops. Furthermore, Möller *et al.* (2008) reported average soil mineral N content in spring. In this case, they found significantly higher soil mineral N content of the digested slurry treated samples (Markadi *et al.*, 2012).

2.5 Effects of anaerobic Digestates in the environment:

a) Ecosystem Engineers

Anyanwu *et al.* (2017) reported that the chemical (nitrogen-containing polycyclic aromatic hydrocarbons (N-PAHs)) had detrimental effects on soil biota. Earthworms exposed to N-PAHs - soils showed weight losses, mortality, and health effects. They also observed - physical damage to the earthworms (breakage of the clitella, body ruptures, body lesions, discoloration and death). However, the effects of digestate on earthworms and other organisms found in the environment have not been explored. This requires urgent research.

b) Plants

Two recent studies in Sweden (Haraldsson, 2008 and Zetterstrom, 2008) showed that common fungal diseases of plants are irreversibly inhibited or killed during mesophilic digestion with a hydraulic retention time of between 25–30 days. The evidence suggested that it is the combination of the conditions in the digester such as; pH level, quantities of volatile fatty acids, the negative effect of ammonium, hydrogen sulfide as well as time and temperature that combine to create the hostile environment in which the spores are unable to survive. This in itself demonstrates the need for caution in making generalizations, since the conditions inside the digester can vary between digesters and between feedstock. Nevertheless, it is reasonable to conclude from the Swedish work that farms with a mesophilic digester would benefit from a significant or destruction of many disease-spreading spores that can affect their crops. The AD, therefore, has the potential to offer real benefits to organic farmers and those wishing to reduce the use of fungicides.

However, some plants do not develop well in digestate amended soils for unknown reasons. Thus phytotoxicity tests are required.

d) Soil Organic Matter Content

Land application of anaerobic digestates may introduce into soils physical, chemical and biological contaminants which may affect their long-term agricultural productivity (Markadi *et al.*, 2012).

d) Soil pH

Fuchs and Schleiss (2008) reported that the digestates enhance the soil pH for about ½ unit. The digestate also enhances the biological activity in the soil. (Jacques *et al.*, 2008)

e) Soil Carbon and Nitrogen

Odlare *et al.* (2008) reported that soil chemical properties hardly change in the short term when the soil is amended with digestates. However, nitrogen leaching has received most of the attention from researchers due to the considerable amount of nitrogen in animal manures and slurries. (Nkoa, 2014).

BENEFITS OF ANAEROBIC DIGESTATE

Bio-fertilizers (such as anaerobic digestates) are environmentally friendly fertilizers that not only prevent damages to natural sources but help to some extent in cleaning the nature of precipitated chemical fertilizers (FAO, 2008). In addition, bio-fertilizers are one of the best modern tools for agriculture and it is a gift of our modern agricultural science. Bio-fertilizers are applied in the agricultural field as a replacement to our chemical fertilizers because they are cost-effective (Owamah *et al.*, 2014).

The term "bio-fertilizer" refers to nutrient supplement inputs for plant growth that are of biological origin. Bio-fertilizers accelerate certain microbial processes in the soil which augment the extent of availability of nutrients in a form easily assimilated by plants and also mobilizing nutritive elements from non-usable form to usable form through biological processes (FAO, 2012). The role of bio-fertilizers in agricultural production assumes special significance, particularly in the present context of expensive chemical fertilizers. Moreover, it provides the farmers with a new strategy which is helpful for achieving the targeted goal of food security by increasing the high production yield of food grains (FAO, 2012).

Digestate(s) serves as an excellent organic fertilizer that can replace inorganic fertilizers and raw manure whilst providing various benefits:

i. Reduction of energy consumption and CO_2 emissions:

Digestate arises naturally as a result of controlled biological decomposition of biodegradable substrates and unlike inorganic fertilizer, it does not require any additional energy in the production process. Moreover, by using digestate as an organic fertilizer for crops and as a soil improver/conditioner, the production of artificial fertilizers, which results in additional costs and emissions, can be reduced (Envirotech, 2017). Studies have shown that around 13 kg CO2eq/ton can be saved when digestate replaces mineral fertilizer. Greater environmental compatibility through effective prevention of land contamination and reduce methane emissions. The treatment of manure by AD helps prevent land contamination.

In many developed countries, manure is spread out on fields without prior treatment against pathogens causing potential biological contamination. The AD process at mesophilic temperatures, typically between 35 and 45°C, greatly reduces the number of plant and animal pathogens within the feedstock and at thermophilic temperatures (above 50°C) even destroys viruses in most cases (Envirotech, 2017).

ii. The reduction of greenhouse gas emissions:

Digesting farm manure reduces nearly 90 % of usual greenhouse gas emissions caused by conventional farm manure storage. When, in turn, organic waste is digested instead of used in landfills, detrimental methane emissions are significantly reduced. Additionally, valuable and scarce nutrients such as phosphorus can be recycled back into the soil, thereby contributing to the circular economy without polluting groundwater through landfill leakage (http://www.envirotech-online.com).

iii. Simple application and greater agricultural output:

When digestate is used to replace raw manure as a fertilizer, plants receive a greater benefit. The reason for this is that unlike the nitrogen in the raw manure, the ammonia within digestate is absorbed immediately by the soil. In this way, it directly contributes to plant growth without adhering to plants or the surface of the ground (http://www.envirotech-online.com). Furthermore, digestate has three other remarkable advantages for the agricultural practice: (1) it does not present the odor nuisance specific to raw manure, therefore providing increased land application options, (2) it makes weed control easier and more efficient for farmers, as the process of anaerobic digestion helps to destroy unwanted weeds and viable plant propagules. (3) It is more homogenous, which makes fertilizer spreading more uniform (Envirotech, 2017).

In the agricultural output, it is important to consider effective nutrient management while applying the product. As in the case of all fertilizers, it is crucial to optimize the spreading season of digestate in order to prevent nutrient leakage. This means that digestate should not be applied during low plant take-up, but instead, it should be stored until the growing season. Furthermore, the application of digestate should be taken into consideration and adjusted according to the types of crops grown, as well as according to the type of soil (http://www.envirotech-online.com).

Benefits of anaerobic Digestate in agriculture:

- Transformation of organic waste to very high-quality fertilizer.
- Improved utilization of nitrogen (by plants) from animal manure.
- Balanced phosphorus/potassium ratio in digestate.
- Homogenous and light fluid slurry.
- AD virtually destroys all weed seeds, thus reducing the need for herbicides and other weed control measures.
- Provides a closed nutrient cycle.
- Also, Treated effluent from the AD is a good animal feed when processed with molasses and grains (Ngumah *et al.*, 2013).

g) Benefits of Digestate in the Environment

- **Odors**

Animal manures and many organic wastes contain volatile organic compounds (e.g. iso-butonic acid, butonic acid, isovaleric acid, and valeric acid, along with at least 80 other compounds) that can produce unpleasant odors. However, digestion significantly reduced concentrations of many of these compounds, such that their potential for giving rise to offensive and lingering odors during storage and spreading was significantly reduced thereafter, the use of appropriate spreading methods can prevent the release of any residual odor (Hansen *et al.,* 2004). For example, injection of digestate (or slurry) into the soil largely eliminates odor and loss of ammonia. It is important, in minimizing the disturbance of the digestate during its transfer from the storage tank to the spreaders, as this can result in a release of odor.

Other benefits of digestate to the environment include:-
- Reduces emission of greenhouse gases (GHG).
- Reduces nitrogen leaching into ground and surface waters.
- Improves hygiene through the reduction of pathogens, worm eggs, and flies.
- Reduces odor by 80%.
- Controlled recycling/reduction of waste.
- Reduces deforestation by providing a renewable alternative to wood fuel and charcoal.
- Biogas burns "cleaner" than wood fuel, kerosene, and undigested bio-waste.

It creates an integrated waste management system that reduces the likelihood of soil and water pollution compared to the disposal of untreated bio-wastes (Ngumah *et al.*, 2013).

2.5 Environmental Risks Associated with Applications of Anaerobic Digestates

Anaerobic digestates is a by-product of anaerobic digestion and it may pose a threat to one or several components of the broader environment. For example, anaerobic digestates can directly impact soils, water bodies, and the atmosphere (Nkoa, 2014). Since the initial feedstocks have been depleted of most of their easily degradable carbon during digestion, nitrous oxide is the only significant GHG that can be potentially be released by anaerobic digestates (Nkoa, 2014). Inappropriate storage or application of anaerobic digestates can lead to gaseous nitrogen emission (ammonia and nitrous oxide) and/or nutrients leaching and runoff into surface and ground waters.

2.5.1 Risks of nutrient pollution

A major environmental concern with the land application of digestates is the potential contamination of surface and ground waters with excess nitrogen and phosphorus. Studies have shown that digestates are richer, in terms of nutrient contents than their respective raw manure (Haraldsen *et al.*, 2011; Möller *et al.*, 2008; Chantigny *et al.*, 2008; Gomez et al., 2007; Loria and Sawyer 2005).

Consequently, environmental issues associated with the production and land applications of manures are equally, perhaps potentially more prominent with anaerobic digestates. Example, issues such as surface and groundwater pollution and eutrophication of water bodies, have been linked to the production and use of manure (Mulla *et al.* 2001; Hubbard and Lowrance 1998; Newton *et al.*, 1994; Odgers 1991).

Nutrient leaching potential following application of anaerobic digestates depends upon factors such as fertilization strategies (e.g. time and methods of application), soil texture (e.g. sandy and clayey soils), topography, precipitations, and cropping systems (Nkoa, 2014).

2.3 Risks of soil contamination

Land application of anaerobic digestates may introduce into soils physical, chemical and biological contaminants (Nkoa, 2014).

2.3.1 Physical contaminants

Physical contaminants are considered to be all the non- or low-digestible materials e.g. plastic, glass, metal scrap, stones, sand, and wood. Such physical impurities are likely to be present in all types of feedstock, but most frequently in household wastes, food waste, garden waste, straw, solid manure, and other solid types of waste. The presence of physical contaminants can cause a negative public perception of digestate as well as aesthetic damage to the environment (IEA, 2017).

2.3.2 Chemical contamination

Chemical contamination of digestate usually comes from human sources such as sewage and includes inorganic materials (e.g. heavy metals) and persistent organic compounds such as polycyclic hydrocarbons (PAHs). Agricultural by-products can contain small quantities of antibiotics, disinfectants, and ammonium (Al Seadi, 2001).

Land application of digestate is not risk-free, since it may result in the incorporation of phytotoxic compounds, pathogens, and heavy metals into the soil (Boydston *et al.*, 2008). With respect to the phytotoxicity of digestates, causal compounds include ammonia (Leege *et al.*, 1997). The high concentrations of some micronutrients such as Cu and Zn in digestates pose a serious concern (Alburquerque *et al.*, 2012) due to the use of pig and cattle slurry as feedstock. Besides heavy metals, micronutrients could be a threat to sustainable agriculture. A recent survey across Europe has disclosed the abundance of micronutrients in all digesters, especially those supplied with wastes like blood, kitchen, and food wastes (Schattauer *et al.*, 2011). For example, Cu and Zn can potentially affect the sustainability of agricultural soils through soil accumulation and interference with the metabolic activities of plants; they can inhibit plant growth once in excess in the soil solution (Ebbs *et al.*, 1997).

Further, manganese is an essential element for plant growth and development. It can, however, be detrimental when available in excess. Long-term applications of digestates onto lands may result in Mn and organic matter accumulation; factors that can cause Mn toxicity, especially in soils with low Mn absorption capacity.

2.3.3 Biological Contamination

The presence of biological contaminants in digestate, such as various pathogens, prions, seeds, and propagules, may result in new routes of pathogen and disease transmission between animals, humans, and the environment (Lukeurst *et al.*, 2010). Thus, strict control of specific feedstock types and of digestate is required (Lukeurst *et al.*, 2010).

Numerous pathogenic bacteria species have been counted in organic wastes used for the anaerobic digestion (Sahlstrom, 2003). The risk is higher when manure is included as a feedstock since several outbreaks of gastroenteritis have been linked to livestock operations (Spencer *et al.*, 2004). Despite the production and hygienization process, the persistence of pathogenic parasite eggs, bacteria, and fungi have been reported in many biogas plants (Slana *et al.*, 2011).

Stabilization of digestates through post-treatment measures such as curing (Drennan et al., 2010) and composting (Tiquia *et al.*, 1996; Smet *et al.*, 1998) significantly reduces the risk they pose on human health and the broad environment.

2.4 Other Risks of Digestate to the Environment

Mismanagement of digestate can cause serious impacts to surface water bodies because of the amount of nitrogen and phosphorus in the digestate. These impacts can lead to contamination of drinking water supplies, eutrophication, the growth of algae, and a decrease in water oxygen levels that negatively affects the survival of aquatic organisms (Global Methane Initiative, 2017).

CHAPTER 3

3.0 CONCLUSION AND RECOMMENDATIONS

3.1 Conclusion

Anaerobic digestate may be rich in macro and micronutrients and can be used in agriculture for the improvement of soil fertility. The fertilizer efficacy of digestates depends upon factors such as the nature of the feedstock, the method of storage and handling, and the method of field applications (Nkoa, 2014).

Microbiological activity of the soil, as well as plant growth, could also be increased by the application of digestate (Makádi *et al.*, 2012).

However, beyond these "classical" application possibilities of digestate, there may be negative impacts of anaerobic digeastate in the environment/agriculture, especially on ecosystem engineers. Thus, studies need to focus on these aspects to help understand the environmental and agricultural effect of soils amended with anaerobic digeastate.

- The fertilizer efficacy of digestates depends on the nature of the feedstock, the method of storage/handling, and the method of field applications.

- There are new promising alternatives for its utilization which means anaerobic digestate is rich in macro and micronutrients and can be used in agriculture for the improvement of soil fertility.

3.2 Recommendation

Therefore, from the prospect and challenges in anaerobic digestate that have been discussed that I recommend as follows:

- Increased knowledge on the effects of digestate on soil nutrient, soil biota, and crop yield is needed.

- Large-scale production and modern packaging of digestate as bio-fertilizer should be encouraged with standard labeling of their nutrient content.

- The digestate should be applied to the soil during the growing season to avoid loss of nutrients.

3.3 Future Research

There is a need to investigate:

- The impacts of digestates on ecosystem engineers (microbes, earthworms, enchytraeids, collembolans, and plants).

- The heavy metals and PAHs contents of the digestate.

- The utilization of anaerobic digestate as solid fuel after drying.

References

Alburquerque, J., de la Fuente, C., Ferre-Costa, A., Carrasco, L, Cegarra J., Abad, M. and Bernal, M. (2012). Assessment of the fertilizer potential of digestates from farm and agro-industrial residues. *Biomass Bioenergy,* 40:181–189.

Amon, T., Amon, B., Kryvoruchko, V., Zollitsch, W., Mayer, K. and Gruber, L. (2007). Biogas production from maize and dairy cattle manure— the influence of biomass composition on the methane yield. *Agricultural Ecosystem Environment,* 118:173–182.

Aniche, P. (2013). Anaerobic Digestion of Waste Business Round Table on Renewable Energy. *Midori Environmental Solutions.*

Anna, C. (2015). Digestate-Maximizing its Value and Use. *Ontario*

Anyanwu, I. N. and Semple, K. T. (2016). Effects of phenanthrene and its nitrogen-heterocyclic analogues aged in soil on the earthworm Eisenia fetida. *Applied Soil Ecology,* 105:151–159.

Anyanwu, I.N., Clifford, O.I. and Semple, K.T. (2017). Effects of Single, Binary and Quinary Mixtures of Phenanthrene and Its N-PAHs on Eisenia fetida in Soil. *Water Air Soil Pollution* 228:105.

Bauer, A., Mayr, H., Hopfner-Sixt, K. and Amon, T. (2009). Detailed monitoring of two biogas plants and mechanical solid-liquid separation of fermentation residues. *Journal of Biotechnology,* 142:56–63.

Bioenergy Association of New Zeeland, (2013). *The production and use of digestate as fertiliser.* Retrieved November 29, 2017, from http://www.bioenergy.org.nz/documents/resource/TG08-theproduction-and-use-of-digestate-as-fertiliser.pdf.

Boydston, R.A., Collins, H. and Vaughn, S. (2008). Response of weeds and ornamental plants to potting soil amended with dried distillers grains. *Horticulture Science* 43:191–195.

Brändli R. C. *et al.*, (2005). Persistent organic pollutants in source-separated compost and its feedstock materials—a review of field studies. *Journal of environmental quality, 34,* 735–760.

Chunlan, M., Yongzhong, F., Xiaojiao, W. and Guangxin, R. (2015). Review of research achievements of biogas from anaerobic digestion. *Renewable and Sustainable Energy Reviews.*

Colleran, E. (2000). Hygienic and sanitation requirements in biogas plants treating animal manures or mixtures of manures and other organic wastes. *Herning Municipal Authorities*, 46:77–86.

Dahihiru, U. H. and Abdulsalam, S. (2017). Development of a bench scale biodigester for the production of bio-fertilizer using cow dung and watermelon peels. *Chemical and Process Engineering Research,* 47:2225-0913.

Digestate Anaerobic Digestion (2017). *Digestate Factsheet.* Retrieved from http://www.biogasinfo.co.uk/about/digestate. Accessed: November, 24 2017.

Drennan, M.F. and Distefano, T.D. (2010). Characterization of the curing process from high solids anaerobic digestion. *Bioresource Technology,* 101:537–544.

Drosg, B., Fuchs, W., Al Seadi, T., Madsen M. and Linke B. (2015). Nutrient Recovery by Biogas Digestate Processing. *IEA Bioenergy.*

Dumitru, M. (2014). Studies concerning the utilisation of digestate in biogas plants. *Studies, 14*(1).

Ebbs, S.D., Kochian, L.V. (1997). Toxicity of zinc and copper to Brassica species: implications for phytoremediation. *Environmental Quality,* 26:776–781.

Engeli, H. Eidelmann, W., Fuchs. J. and Rottermann M. (1993). Survival of plant pathogens and weed seeds during anaerobic digestion. *Water Science and Technology,* 27:69–76.

Fang, H.H., and Liu H. (2002). Effect of pH on hydrogen production from glucose by a mixed culture. *Bioresource Technology*, 82:87–93.

Fuchs, W. and Drosg, B. (2013). Assessment of the state of the art of technologies for the processing of digestate residue from anaerobic digesters. *Water Science & Technology*, 67(9).

Global methane. (2018). Retrieved from http://www.globalmethane.org. Accessed 28 December 2017.

Gomez, X., Cuetos, M.J., Garcia, A.I. and Moran, A. (2007). An evaluation of stability by thermogravimetric analysis of digestate obtained from different biowastes. *Hazard Mater* 149:97–105.

Gutser, R., Ebertseder, T., Weber, A., and Schraml, M. (2005). Short-term and residual availability of nitrogen after long-term application of organic fertilizers on arable land. *Plant Nutrient Soil Science,* 168:439–446.

Hansen, M.N., Birkmose, T., Mortensen, B., Haraldsen T.K., Andersen U., Krogstad, T. and Sørheim R. (2011). Liquid digestate from anaerobic treatment of source-separated household waste as fertilizer to barley. *Waste Management Resources*, 29:1271–1276.

Holm-Nielson, J.B., Al Seadi, T. and Oleskowicz-popiel (2009). The future of anaerobic digestion and biogas utilization. *Bioresource Technology*, *100*, 5478-5484.

International Energy Agency (2005). *Biogas Production and Utilisation.* Retrieved November 16, 2017, from http://www.iea-biogas.net.

International Energy Agency (2010). *Utilisation of digestate from biogas plants as bio-fertiliser.* Retrieved November 16, 2017, from http://www.iea-biogas.net.

Jacques, G., Urs B., Alfred B., Jochen, M. and Konrad, S., (2008). Effects of digestate on the environment and on plant production - results of a research project. *FiBL*.

Kajsa, R. (2015). *Quality and function of anaerobic digestion residues* (unpublished doctoral thesis), Swedish University of Agricultural Sciences, Uppsala.

Kratzeisen, M., Starcevic, N., Martinov, M., Maurer, C. and Müller, J. (2010). Applicability of biogas digestate as solid fuel. *Fuel*, 89:2544–2548.

Leege, P.B. and Thompson, W.H. (1997). Test methods for the examination of composting and compost. *The US Composting Council*.

Liu, D. and Christians, N. (1994). Isolation and identification of root-inhibiting compounds from corn gluten hydrolysate. *Plant Growth Regulation*, 13:227–230.

Lukehurst, C., Frost, T. and Al Seadi, T. (2010). Utilization of digestate from biogas plants as biofertilizer. *IEA Bioenergy*, 37:4-17.

Lukehurst, C.T. (2010). *The use of digestate in the UK.* Retrieved from http://www.iea-biogas.net/publications/Workshops/Copenhagen. Accessed 25 November 2017.

Luste, S. and Luostarinen, S. (2010). Anaerobic co-digestion of meat-processing by-products and sewage sludge: Effect of hygienization and organic load rate. *Bio-resource Technology* 101:2657–2664.

Makádi, M., Tomócsik, A. and Orosz, V. (2012). Digestate: A new nutrient source – Review *Research Institute of Nyíregyháza,* 14:295-307.

Mejnertsen P. and Møller, H.B. (2011). Nitrogen fertilizer value of digestates from anaerobic digestion of animal manures and crops. In P. Sørensen (Eds.), Proceeding in Nordiac Association of Agricultural Scientist Seminar, Denmark: Department of Agroecology, Aarhus University.

Møller, H. B., Lund, I. and Sommer S. G. (2000). Solid-liquid separation of livestock slurry: efficiency and cost. *Bioresource technology*, 74:223–229.

Möller, K., Stinner, W., Deuker, A. and Leithold, G. (2008). Effects of different manuring systems with and without biogas digestion on the nitrogen cycle and crop yield in mixed organic farming systems. *Nutrient Cycle Agroecosystem,* 82:209–232.

Möller. K. and Müller, T. (2012). Effects of anaerobic digestion on digestate nutrient availability and crop growth: a review. *Engineering Life Science*, 12:242–257.

Mulla, D.J., Birr, A.S., Randall, G., Moncerief, J., Schmitt, M., Sekely, A., and Kerre, E. (2001). Impacts of animal agriculture on water quality: technical work paper. *Environmental Quality.*

Ngumah, C., Ogbulie, J., Orji, J. and Amadi, E. (2013). Potential of Organic Waste for Biogas and Biofertilizer Production in Nigeria. *Environmental Research, Engineering and Management,* 63:60-66.

Ngumah, C., Ogbulie, J., Orji, J. and Amadi, E. (2013). Potential of Organic Waste for Biogas and Biofertilizer Production in Nigeria. *Environmental Engineering 7:*110-116.

Nicholas, K., Vera P., Elena, Z., Viktor, K., Svetla, M. and Plamen, I. (2015). Impacts of Biogas Digestate on Crop Production and the Environment: A Bulgarian Case Study. *American Journal of Environmental Sciences* 2:81-89.

Nkoa R. (2014). Agricultural benefits and environmental risks of soil fertilization with anaerobic digestates: A review. *Springer.*

Oenema, O. and Tamminga, S. (2005). Nitrogen in global animal production and management options for improving nitrogen use efficiency. *Life Sciences*, 48:871–887.

Parawira, W., Murto, M., Zvauya, R. and Mattiasson, B. (2004). Anaerobic batch digestion of solid potato waste alone and in combination with sugar beet leaves. *Renew Energy* 29:1811–1823.

Provenzano, M.R., Iannuzi, G., Fabbri, C. and Senesi, N. (2011). Qualitative characterization and differentiation of digestates from different bio-wastes using FTIR and fluorescence spectroscopies. *Environmental Protection,* 2:83–89.

Rachel, M., Alfonso, J.L., Ben, H., Ian, C. D., John, Q., Ben, S., Farid, A. and Kirk T. S. (2017). Digestate and ash as alternatives to conventional fertilizers: Benefits and threats to soil biota. In H. Lois & O. Nick (Eds.), Proceedings of the EGU General Assembly conference, Chester, United Kingdom.

Rehl, T. and Müller, J. (2011). Life cycle assessment of biogas digestate processing technologies," Resources, Conservation and Recycling, 56:92–104.

SahlstromL, M. (2003). A review of survival of pathogenic bacteria inorganic waste used in biogas plants. *Bio-resource Technology*, 87:161–166.

Schattauer, A., Abdoun, E., Weiland, P., Plöchl, M. and Heiermann, M. (2011). The abundance of trace elements in demonstration biogas plants.

Schnurer, A. and Schnurer, J. (2006). Fungal survival during anaerobic digestion of organic household waste. *Waste Management*, 26:1205–1211.

Slana, I., Pribylova, R., Kralova, A. and Pavlik, I. (2011). Persistence of Mycobacteriumavium subsp. Para-tuberculosis at a farm-scale biogas plant supplied with manure from paratuberculosis-affected dairy cattle. *Applied Environmental Microbiology* 77:3115–3119.

Spencer, J.L. and Guan, J. (2004). Public health implications related to spread of pathogens in manure from livestock and poultry operations. *Public health microbiology,* 28:503–515.

Svensson, K., Odlare, M. and Pell, M., (2004). The fertilizing effect of compost and biogas residues from source separated household waste. *Agricultural Science,* 142:461–467.

Szűcs, B., Simon, M. and Füleky, G. (2006). Anaerobic pre-treatment effects on the aerobic degradability of wastewater sludge. *Proceedings of the International Conference ORBIT,* 2:425-434.

Tambone, F., Genevini, P., D'Imporzano, G. and Adani, F. (2013). Assessing amendment properties of digestate by studying the organic matter composition and the degree of biological stability during the anaerobic digestion of the organic fraction of MSW," Bio-resource technology, 100(12):3140–3142.

Tambone, F., Scaglia, B., D'Imporzano, G., Schievano, A., Salati, V. and Adani, F. (2010). Assessing amendment and fertilizing properties of digestates from anaerobic digestion through a comparative study with digested sludge and compost. Chemosphere 81:577–583.

Tasneem, A. (2012). *A Brief History of Anaerobic Digestion and Biogas*.

Tiquia, S.M., Tam, N.F. and Hodgkiss, I.J. (1996). Effects of composting on phytotoxicity of spent pig manure sawdust litter. *Environmental Pollution,* 93:249–256.

Using quality anaerobic digestate to benefit crops (2012). Retrieved December 29, 2017, from http://www.wrap.org.uk/agriculture.

Vilanova, P., Plana and Noche, B. (2016). A review of the current digestate distribution models: storage and transport. *Wit Conferences*.

Weiland P. (2010). Biogas production: current state and perspectives. *Applied microbiology and biotechnology,* 85:849–860.

Wellinger, A., Murphy, J. D. and Baxter, D. (2013). The biogas handbook: science, production and applications: *Elsevier*.

Wulf, S., Maeting, M. and Clemens, J. (2002). Environmental Technology. Retrieved from http://www.envirotech-online.com/article/water-wastewater/17/susanna-litmanen-franz-kirchmeyr/the-use-of digestate an organic fertilizer/1593. Accessed: 2 December, 2017.

Wulf, S., Maeting, M., Clemens, J. (2002). Application technique and slurry-fermentation effects on ammonia, nitrous oxide, and methane emissions after spreading: I. Ammonia volatilization. *Environmental Quality, 31:*1789–1794.

www.ingramcontent.com/pod-product-compliance
Lightning Source LLC
Chambersburg PA
CBHW081059240526
45465CB00025B/2770